ENGLISH HOUSES
AN ESTATE AGENT'S COMPANION

Pictures, Glossary and Other Matters of Interest

2004

A division of Reed Business Information

Estates Gazette
1 Procter Street, London WC1V 6EU

ISBN 0 7282 0453 3

Typeset by Amy Boyle, Rochester, Kent
Printed by Bell & Bain Ltd., Glasgow

To the people of Saffron Walden and the district, with whom I have had the privilege of being associated – friends, acquaintances, principals, partners, staff, business colleagues, competitors, solicitors and, not least, all those clients who entrusted me with their instructions.

Foreword

Essential reading for all thinking estate agents dealing with town and country properties, but more importantly an introduction to the 'man (or woman) in the street' who, following a perusal of this book, will certainly be more aware of the delights of English houses.

An exceptionally easy to read book containing a wealth of information relating to property, in addition to briefly describing the important features of a great variety of houses to be seen in the streets of England.

This book is a tribute to one of the most enthusiastic and knowledgeable estate agents with over 50 years' experience and dedication to the property profession. It is doubtful whether there has been any Estate agent with a similar depth of knowledge of properties of previous centuries.

I would recommend every interested agent, and others, to read, retain and refer to this excellent book whenever necessary.

Tony Mullucks, FRICS
Mullucks Wells

Contents

Introduction

This is a book for browsing, and a reference book, and is really a later edition of "English Houses" published by the *Estates Gazette* in 1979.

Here is a simple book, principally about houses but including information on matters related to property generally. It is intended to be of interest to those involved in the property profession and to the layperson alike.

The sale of a house is a transaction in which most of us become involved at some time. All sorts of matters crop up in property transactions and these notes should provide some useful information, helpful to both professional and layperson.

The architectural notes are quite superficial but seek to inform about the age period and construction of English houses. Many of the dates are "guesstimates" – it is often not possible to be precise. The salient external architectural features are described and the illustrated glossary should cover most common features not shown in the photographs. Not every style of house is covered, but there is a good general selection of English houses of most periods.

Over the centuries the generations have created different styles, each with their own particular features. Succeeding generations have altered original buildings to such an extent that many houses in this country are a hotchpotch of several periods. Original buildings that seem "of a piece" may have been built in several stages.

Many of the illustrations show houses of early dates refaced in later centuries and it should be borne in mind that a very late and ordinary looking house may conceal internal features of great age and interest.

Some believe that recent planning restrictions have tended to curtail further development so that it may be that the evolution of the English House is at an end. Perhaps for the guidance of all involved in restoration work, both restorers and planners, the words of William Morris are worth quoting. Referring to old buildings, he said "These do not belong to us only, they belonged to our forefathers and will belong to our descendants unless we play them false. They are not in any sense our property to do as we like with them. We are only trustees for those who come after us."

Bringing to the reader's attention the excellent design of some of England's everyday features will lead to a general appreciation of things that may otherwise be considered insignificant, or ordinary, and taken for granted.

So many fine buildings and architectural features go unnoticed. Hopefully the following pages will lead to a greater awareness and appreciation of these things. For those with limited time who seek examples of fine buildings of all periods by great English architects, a visit to the Colleges of Cambridge and Oxford will be a worthwhile experience, particularly if the visitor is armed with the appropriate volume of Nikolaus Pevsner's *The Buildings of England*.

The location of the properties described is intentionally omitted. The design features are generally common to all areas of the United Kingdom even though the materials differ.

The Periods

The dates of the periods are not hard and fast, and vary in different publications. Within reason the following are usually accepted:

		THE SOVEREIGN	
TUDOR	1485–1560	Henry VII	1485–1509
		Henry VIII	1509–1547
		Edward VI	1547–1553
		Mary	1553–1558
ELIZABETHAN	1558–1603	Elizabeth I	1558–1603
JACOBEAN OR EARLY STUART	1603–1649	James I	1603–1625
		Charles I	1625–1649
CROMWELLIAN	1649–1660	Commonwealth	
CAROLEAN OR LATE STUART	1660–1689	Charles II	1660–1685
		James II	1685–1689
WILLIAM AND MARY	1689–1702	William III and Mary	1689–1702
QUEEN ANNE	1702–1714	Anne	1702–1714
GEORGIAN EARLY 1720–1750 LATE 1750–1800	1714–1800	George I	1714–1727
		GEORGE II	1727–1760
		George III	1760–1820
REGENCY AND LATE GEORGIAN	1800–1830	George III and George IV	1820–1830
EARLY VICTORIAN	1830–1837	William IV	1830–1837
VICTORIAN	1837–1901	Victoria	1837–1901
EDWARDIAN	1901–1910	Edward VII	1901–1910

The Medieval period is taken to end at the Battle of Bosworth in 1485. It is not included in the table above, but the origins of many houses date from the 15th century and before.

Gothic denotes the 12th to 16th centuries' style of architecture in churches and other buildings of those periods. The periods were designated by the architect Thomas Rickman in 1812, as follows:

Norman   1066–1190
Early English   1190–1300
Decorated   1300–1370
Perpendicular   1370–1500

Gothic styles were revived in the 18th and 19th centuries and led to such High Victorian Gothick buildings as the Palace of Westminster, St Pancras Hotel and station, as well as many churches, country and town houses.

Gothick (with a "k") denotes the early imitation of the Gothic style of the middle ages, circa 1720–1840.

The Orders

The architecture of ancient Greece and Rome is known as Classical Architecture. The forms and decorations were revived in Italy in the 15th and 16th centuries and were fashionable in Britain from the late 16th century onwards.

The columns evolved by the Greeks and Romans, together with the lintels over them, are called the "Orders" and have greatly influenced British architecture.

Many of the doorcases of houses since the 17th century are based on these orders.

Here are three main types of orders, consisting of columns with their capitals and bases.

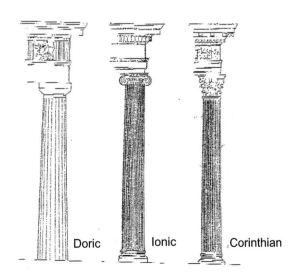

Doric   Ionic   Corinthian

The Arches

The shapes of the arches above windows and doors are referred to as follows:

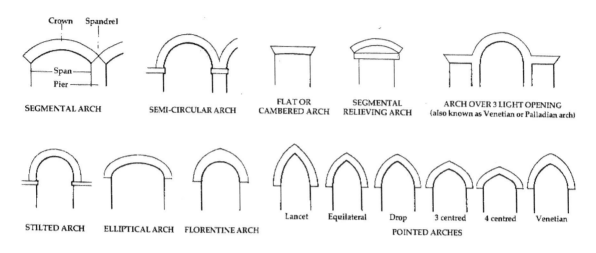

The Roofs

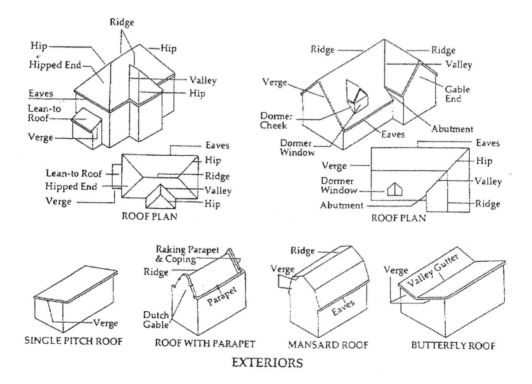

EXTERIORS

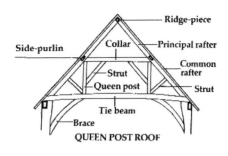

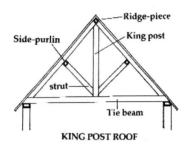

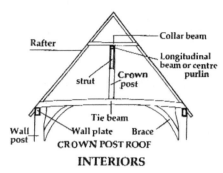

INTERIORS

Building Materials

Main walls

The type of building materials used for the walls depends upon the type of construction.

The two structural systems for houses are:

1. Solid construction in which the loads of roof and floors are carried to foundation by walls.

2. Frame construction in which the loads of roof and floors are carried by a timber frame which in turn is covered for weather protection.

In the former system, the materials used include solid forms, for example brick, flint, building block, and in the latter the frame, which is of timber, has a protective covering or infill of wattle and daub, lath and plaster, brick nogging or weatherboard.

The principal building materials are as follows:

Ashlar
Hewn stone, used for facing to give a high quality finish.

Brick
The universal building material in England and Wales by mid Victorian times. The pattern in which the bricks is laid is referred to as bonding.

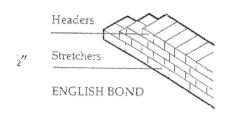

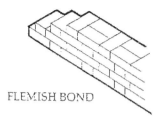

Headers

2"   Stretchers

ENGLISH BOND   FLEMISH BOND

Bricks have varied in size through the centuries. Here is a guide to their dates:

GREAT BRICKS	1200–1520	12"	6"	1¾" to 2¾"	
FLEMISH BRICKS	13th C–1571	8"	9¾"	3¾" to 4¾"	1¾" 2½"
STATUTE BRICKS	1571	9"	4½"	2½"	
(regulated by law)					
MODERN BRICKS	18th C	8½"	4"	2¼"	
		8¾"	4"	2⅝" or 2⅞"	

Individual kilns (for example on landed estates) had their own peculiar dimensions.

Brick nogging
Brick filling, sometimes in herringbone pattern, in between timber studs.

Cob
Mud mixed with water and wheat straw, and perhaps lime, then plastered over.

Clay bat
Clay mixed with straw and shaped in rectangular blocks.

Clunch
The usual name for blocks from a hard seam of chalk.

Flint
Flints picked from the fields or quarried, are irregular in shape and the exposed ends in a wall are often round topped. If knapped they are square topped and smooth.

Half timber
Timber framework, usually oak, with the spaces in between the vertical studs filled with (1) plaster on wooden laths or (2) wattle and daub, a form of similar infilling in which clay, dung, horsehair, etc. was daubed onto interwoven wattles. Often the whole was plastered over and patterned. The patterning is called pargetting or pargework.

Stone
Stone from the fields or quarry, depending on the particular area of the country, was used for main walls.

Stucco
A form of cement render used to cover brick or stone.

Vertical tiles or slates
Tiles or slates hung on laths (also called hanging tiles or slates).

Weatherboarding (or clapboard)
Long timber horizontal boards nailed to the framework and tarred or painted.

Roofing Materials

Thatch
The principal thatching materials are straw and reed. Reed is grown in Norfolk, and there are other sources. There are three forms of thatch: Norfolk Reed (or Wet Reed), Long Wheat Straw and Combed Wheat Reed. (See page 8.)

Slate
Slate comes from quarries mainly in Wales, the West Country and the Lake District.
It began to come into use throughout the country in the late 18th century.
Welsh Slate is used in thin slabs of uniform thickness and size. Other slates are more irregular.
In recent years much Spanish slate has been imported.
Stone Tiles are quarried and are to be found in the North and West of the country.

Plain Tiles
Clay tiles 10 ½" 6 ½" ½" were standardised in 1477 and are still manufactured in this size.
Tiles are laid on battens originally held in position by wooden pegs (peg tiles) and later by nibs moulded in the tile. Clay tiles are still manufactured as well as concrete tiles.

Pan Tiles
Shaped tiles moulded in an "S" curve, when laid, slightly overlap each other vertically.

Shingles
Rectangular wood tiles, originally of oak but now generally of western red cedar.

Asbestos Tiles
Rectangular tiles of asbestos cement.

Thatch

Norfolk Reed

Combed Wheat Straw

Long Wheat Straw

Stones

(Extracted from *Stones of Britain*, by B.C.G. Shore)
Used for buildings in the vicinity of quarries, but elsewhere as well. Portland stone (from Dorset), for example was for 300 years or more the principal building stone.

Slates   North Wales (now Spanish are used).
Ashburton Marble   Devonshire

Bargate	Godalming
Barnack	Northamptonshire
Bath	Bath district
Beckfoot Granite	Cumberland
Blue Lias	Somerset and Dorset
Broughton Moor	Lake District
Carstone	Norfolk
Charlbury	Oxfordshire
Chert	Kent
Chilmark	Wiltshire
Clipsham	Lincolnshire
Clunch	Cambridgeshire
Collyweston	Lincolnshire
Corsham Down	Somerset
Cotswold Dale	Gloucestershire
Darley Dale	Derbyshire
Dartmoor Granite	Devonshire
Downton	Shropshire
Folkestone	Kent
Gatton	Surrey
Ham Hill	Somerset
Happaway "Marble"	Devonshire
Headington	Oxfordshire
Honister Stone	Lake District
Hornton	Oxfordshire
Horton Stone	Yorkshire
Howley Park	Yorkshire
Ikley Moor Stone	Yorkshire
Kentish Ragstone	Kent
Kenton	Near Newcastle
Ketton	Stamford, Lincolnshire.
Kitley Green Marble	Devonshire
Lamorna Granite	Cornwall
Leckhampton	Gloucestershire
Mansfield	Nottinghamshire
Marchalee Elm Stone	Somerset
Monks Park	Bath
Mountsorrel Granite	Leicestershire
Nailsworth	Cotswolds
Ogwell Marble	South Devonshire
Oldbury	Kent
Park Nook	Yorkshire
Pea Grit	Near Cheltenham
Pennant	Bristol
Petworth Marble	Sussex

Portland Stone	Dorset
Prudham Stone	Northumberland
Purbeck Cliff Stone	Dorset
Reigate Firestone	Surrey
Robin Hood	Yorkshire
St. Boniface Stone	Isle of Wight
Sarcen	Wiltshire
Shap Fell Granite	Westmoreland
Storeton	Cheshire
Taynton	Oxfordshire
Tisbury	Wiltshire
Totternhoe	Bedfordshire
Trevor	Penmaenmawr, North Wales.
Wealden	Sussex and Kent
Weldon	Northamptonshire
Wellfield	Yorkshire
Wenlock	Shropshire
Wheatley	Near Oxford
Woodburn Stone	Northumberland
Woolton Stone	South Lancashire

Glossary

Architrave
A moulded enrichment to the jambs and head of a doorway or window opening.

Baluster
A post supporting a hand rail, usually part of a series called a balustrade.

Barge-board
A board, sometimes carved, fixed to the edge of a gabled roof, a short distance from the face of the walls.

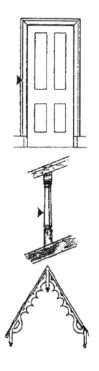

Bay Window
Projecting window with angles.

Bow Window
Projecting convex window.

Brace
In roof construction, a timber inserted to strengthen the framing of a truss.

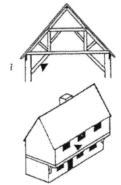

Bressumer
A beam forming the direct support of an upper wall in timber framing, similar to a lintel.

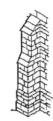

Buttress
Projecting masonry built against a wall to give additional strength.

Canopy
A projection or hood over a door or window.

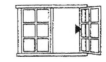

Casement
The opening part of a window.

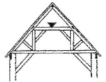

Collar-beam
A horizontal beam serving to tie a pair of rafters together some distance above the wall plate level.

Corbel
A projecting stone or piece of timber for support; or oversailing courses of masonry.

Crow-stepped
A term applied to gables, the coping of which rises in a series of steps.

Cruck
Pair of large curved timbers carrying ridge of building direct from ground, used throughout the MIddle Ages.

Cupola
Strictly a dome but more usually describing a domed turret crowning a roof.

Diapering
In brickwork, burnt or glazed bricks set in a criss cross pattern.

Dormer Window
A vertical window on the slope of a roof and having a roof of its own.

Eaves
The under part of a sloping roof overhanging a wall.

Entablature
Classical horizontal feature joining tops of columns or crowning wall face and comprising architrave, frieze and cornice.

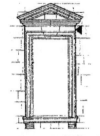

Fan Light
Small window, often semi-circular set in head of door opening.

Fascia
A plain or moulded board covering either the rafter feet at the eaves or the plate of a projecting upper story.

Finial
A formal bunch of foliage or similar ornament at the top of a pinnacle, gable, canopy, etc.

Gable
The wall at the end of a ridged roof, generally triangular, sometimes semi-circular.

Dutch-Gable
A gable with multi-curved sides.

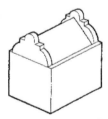

Hammer-beams
Horizontal brackets of a roof projecting at the wall plate level, and resembling the two ends of a tie beam with its middle post cut away, supported by braces or struts.

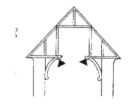

Hood mould (or moulded lintel)
A projecting moulding on the face of a wall above an arch, doorway, or window. Sometimes it follows the form of an arch and sometimes it is square in outline.

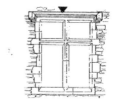

House Classifications
1. Hall and cellar type. Hall on first floor, rooms beneath generally vaulted.

2. H Type. Hall between projecting wings one containing living rooms, the other the offices. This is a common form of a medieval house, employed with variations down to the seventeenth century.

3. L Type. Hall and one wing. Generally small houses.

4. E Type. Hall with two wings and a middle porch, usually sixteenth century and seventeenth century.

5. Half H type. A variation of the E type without the middle porch.

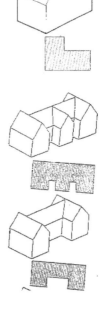

6. Courtyard type. House built round a court, sometimes only three ranges of building with or without an enclosing wall and gateway on the fourth side.

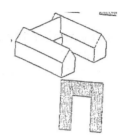

7. Central chimney type. Rectangular plan.

Jettied Storey
Projecting upper storey.

Jambs
The sides of an archway, doorway, fire place, window or other openings.

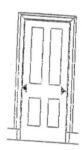

Key Stone
The middle stone in an arch.

Kneeler
Stone at the foot of a gable.

Lancet
A long narrow window with a pointed head typical of the 13th century.

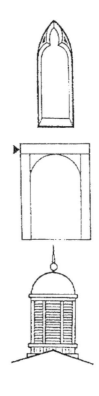

Lintel
The horizontal beam or stone bridging an opening.

Louvre or Luffer
A lantern-like structure surmounting the roof of a hall or other building with openings for ventilation or the escape of smoke.

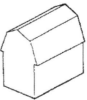

Mansard Roof
A form of roof having a break in the slope, the lower part being steeper than the upper.

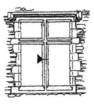

Mullion
An upright between two lights of a window.

Modillions
Brackets under the cornice in Classical architecture.

Muntin
The intermediate uprights on the framing of a door, screen or panel, stopped by the rails.

Oriel Window
A projecting bay window carried upon corbels or brackets.

Oversailing Courses
A number of brick or stone courses, each course projecting beyond the one below it.

Pargetting
Ornamental plasterwork on the exterior of a building.

Pediment
A low pitched gable in Classical Architecture above a door, window or porch.

Pilaster
A shallow pier or column attached to a wall.

Plinth
The projecting base of a wall.

Queen Posts
A pair of vertical posts in a roof truss equidistant from the middle line.

Quoins
The dressed stones or bricks at the angle of a building (pronounced "coins").

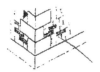

Riser
The vertical board of a step from tread to tread.

Shaft
The part of a chimney above the roof, in particular, the separate stalk terminating each flue. A small column.

Sash Window
A window having movable sash or sashes (usually vertically sliding) the sashes being the frame holding the glass.

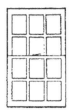

Soffit
The underside of a staircase, lintel, cornice, arch or canopy, the underside of a fixed beam or eaves.

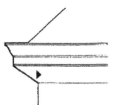

Spandrel
The space between the curve of an arch and the right angle formed by the jamb and the lintel. The space between a curved brace and a tie team, and any similar triangular form.

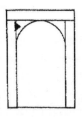

Strut
A timber forming a sloping support to a beam.

Style
The vertical members of a frame into which are tenoned the ends of the rails or horizontal members.

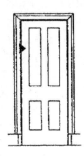

String
The sloping side piece enclosing or supporting the steps of a staircase.

Stud
The vertical post in a partition.

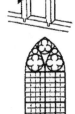

Tracery
The ornamental work in the head of a window, screen or panel formed by the curving and interlacing of bars of stone or wood, and grouped together usually over two or more lights or bays.

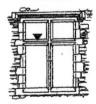

Transom
A horizontal bar of stone or wood across a window opening, doorway or panel.

Truss

A number of timbers framed together to bridge a space or form a bracket, to be self supporting and to carry other timbers. In roof construction there is the Crown Post Truss, the King Post Truss and the Queen Post Truss for example.

Vaulting

An arched ceiling or roof of stone or brick, sometimes imitated in wood.

Wall Plate

A timber laid lengthwise on the wall to receive the rafters or studs.

The Great Changes in Domestic Houses of the 18th Century

Why only note the changes of the 18th century? The most significant changes in architectural styles occurred at that time.

Although designs of houses have evolved over the years and one style has succeeded another, until the 18th century change was gradual and often unnoticeable. The 18th century was a time of immense change and development.

Many would say that the finest domestic buildings date from 1700 or thereabouts, the beginning of the Queen Anne and Georgian periods.

It was a time when outstanding classical features of the larger country houses, the stately homes of the 17th century, influenced and became part of the designs of the smaller town and country houses.

Indeed, later in the 18th century, so fashionable were these classical features that a vast number of village and town houses, large and small, detached and terraced, were given "facelifts", the timber framed gabled elevations of the Medieval and Tudor periods being concealed behind new classical facades.

For years the buildings in the streets of south and east England were covered in posts and barrelled scaffolding while craftsmen replaced gables with pitched roofs, plastered over beams, covered or removed mullions and casements and constructed in their stead the latest sash windows and, with great flair, built the crowning glory of so many house fronts, the elegant door cases with pilasters and hoods.

One has only to walk down the older streets of any town to imagine the changes to the fronts of houses which were wrought 200 years ago – over the channel the French Revolution, here the Domestic Buildings Revolution!

In the late 17th century, English architects studied in Italy and, particularly influenced by the work of Andrea Palladio (1518–80), returned to introduce to this country the great classical revival, incorporating in their designs the architectural features of ancient Rome (and later Greece). These forms and decorations were revived in Italy in the 15th and 16th centuries and became fashionable in Britain from the 16th century onwards.

Here was a sort of architectural Common Market.

The columns evolved by the Greeks and Romans, together with the lintels over them, are called Orders and greatly influenced British architecture from the 17th century. The parapets, door cases, window arches and pediments of the 18th century houses are all the products of the Renaissance, the re-birth of the ideas of Greece and Rome which caused a fundamental change in the appearance of the towns in England – and Europe.

Today, no style of house is more sought after than that of the Queen Anne and Georgian periods – their proportions and elegance, their relation to human form, their Englishness all give them great appeal.

However, what is perceived to be a Queen Anne or Georgian house is unlikely to be "of a piece". Some are purpose built so to speak, but most are a hotchpotch of periods: Medieval; Tudor with 18th century additions; Queen Anne – Georgian fronts outside; panelling, cornices, Adam fireplaces, panelled doors, cased beams inside. Each generation, indeed each owner and occupier, leaves his or her mark, some important, some less so, some good, some bad – each is a custodian only – there's a romance in living in an old house, to be one of a long, long, line.

With the help of text books for the guidance of provincial builders, with plans of the great houses written by noted architects and with pattern books of door cases and windows and other joinery, local builders were able to design houses which were aesthetically pleasing and incorporated the classical features of the past.

An inspection of several houses in the same village will often reveal features so similar that they must have been designed, planned and crafted by the same men.

The most important feature of the house of the 18th century was symmetry. A central entrance, sometimes with steps and railings, balanced windows on either side. Here was dignity. For the exterior, brick was preferred although local materials, stone and lath and plaster, continued in use.

Plans were usually square or rectangular, with rooms on either side of a central hall, the latter often with an important staircase, and, even if the proportions only allowed for a staircase to be concealed, it was often of fine design and craftsmanship.

A frequent feature of the early 18th century doorways was the projecting canopy, either flat or hooded.

The early classical houses had eaves cornices, but later the eaves and roof were concealed behind parapets. Other important and common features were red brick in Flemish bond: that is alternate headers and stretchers. The quoins or angles of the buildings were pronounced by rusticated stonework. There were sash windows with pronounced arches and jambs, often dormers with hipped, gabled or arched roofs peeping above parapets. Later in the 18th century, the facades became plainer with large bricks, windows generally square headed instead of curved, external doorways had triangular pediments, columns and fanlights.

To this day a not uncommon feature in the front of houses is bricked up windows – and why? In 1697 a window tax was levied according to the number of windows, a tax not repealed until 1851.

The rear elevations of refronted houses were often left unaltered and retained an irregular Tudor appearance: gables, casements and beams.

Chats on English Houses

Medieval/Tudor

A very good example of its period. The left hand end, the south, is the earliest part: a timber framed house of circa 1470, cased in brick about 1550.

The principal house was built in the early 1500s in Tudor brick with stone dressings and diapered brickwork (burnt bricks set in a criss-cross pattern). The parapets are castellated.

The house is an interesting example of Tudor and earlier work; the oriel window (the projecting bay) is especially fine.

The tower at the right hand end contains the staircase.

The lantern, or cupola, is a most elegant feature. The arched porch is of two storeys – a four centred archway.

The house is moated, as were many country houses. Shakespeare refers in Richard II to "a moat, defensive to a house" – so one purpose of a moat was for security; they were also status symbols.

Medieval/Tudor/Georgian

A Wealden house, so called since it is built in the style of houses in the Weald of Kent. A handsome farmhouse of timber frame and tiled roof. It was built in the 15th century. In the 18th century the single storey wing on the right was added.

The upper storey is jettied at each end of the front and there is a continuous eave across the middle bay supported by curved braces, the features of a Wealden house. There is a 17th century chimney stack with diagonal shafts and a single dormer window, 19th century panelled doors and casement windows. The side wing is supported by a brick buttress.

Concealed herein is an octagonal king post and other notable period features.

Medieval/Tudor

A restored late 15th century house which has retained the mullioned windows and the dragon post. The end jetty has been covered in, but the dragon post remains as an interesting example. It is set diagonally, projecting at the corner to support the joists of a jetty on two adjacent sides, and so called as a corruption of diagonal.

A Medieval town house with a fine dragon post with carved decoration. This post can be seen more clearly to support the two adjacent jetties. In the right hand elevation can be seen the original doorway and (perhaps a shop) window.

Medieval/Tudor

This ancient terrace displays many succeeding features. Built probably independently in the second half of the 15th century, the four buildings have been altered over the centuries, reflecting the change in styles.

The arrangement of the windows (referred to by planners as fenestration) reveals the changing styles. First there were small mullioned windows (top right) with wooden glassless frames and closed with wooden sliding shutters. In the 16th century the availability of glass in small panes led to the development of casement windows with leaded lights (usually small diamond shaped panes). This building shows a variety of windows of many periods, including a nineteenth century sash and recent neo-Georgian bows.

The roof at the left hand end is hipped and gabled. The two upper storeys are jettied.

The plastered cottage on the left has a late eighteenth century front, though it is in fact of similar period. The front elevation of this terrace was plastered until the 1930s when fashion dictated that the timber frame should be revealed.

Early in the 20th century it became fashionable to strip the protective plaster from timber framed buildings, revealing the beamed structure. A bad move suggest many experts. Often the timbers are of poor quality and were never meant to be seen.

Late Medieval or Tudor

A good example of a "T" plan house, which like the "H" plan was a building shape used from the 13th to the 17th centuries. It is difficult to date except by expert examination of the timber details and joints.

Notice the jettied storey and the large chimney.

The lean to addition is a characteristic feature of small village houses.

The building still preserves the early hall house plan with a two storied cross wing. The dormer windows were probably added in the 18th century to give the main block another storey.

Late Medieval/Tudor/Elizabethan

A good example of a half timbered house with herringbone brickwork between the studs, a form of construction often described as brick nogging.

There is elaborate wood carving below the oriel window, carved spandrels to the door and curved braces below the projecting storey.

Notice the eaves carried straight across the centre hall on arched braces – a feature common in Kentish houses of this time (hence the term "Wealden" for these houses). The coved eaves are Elizabethan.

What a story this house could tell!

Medieval/Tudor/Victorian

This building, once an inn, was constructed in different periods from the 14th to the 16th century, and much altered in the 17th, 18th and 19th centuries.

Here are moulded bressummers, barge-boards, projecting upper storeys and, most importantly, some impressive parge work dated 1676. Plaster birds, foliage, giants and patterns have been skilfully crafted.

Tudor

A village house, perhaps of late 15th century origin, Medieval if before 1485.

The left hand end was added, perhaps a hundred years later.

The closer the oak studs (vertical timbers) the earlier the date, as a rule. As the oak became scarcer the studs became wider apart.

The projecting upper storey running the entire length of the building is called a "long jetty".

Casement windows of the 19th century (but perhaps renewed) have replaced mullions. The great chimney suggests large open fireplaces.

Tudor/Georgian/Victorian

An interesting configuration. A terrace of individual dwellings, all probably of Tudor origin.

On the left, the gabled jetty with a perhaps 17th century bargeboard and an early 19th century bay.

The chimney could not have been planned, but just grew from the principal stack of the 16th century with later shafts – what a feature.

The original mullioned windows, which would have been shuttered, are still exposed. The sash windows are of the 19th century.

Medieval

This shows cruck construction, where a pair of curved timbers rises from the ground to the ridge of a building. (One of the pair in this picture is concealed behind a neighbouring house). Note the rough stone construction to the front, and at the side brickwork and stone between the studs, and a stone tiled roof and a dormer with hipped roof.

A massive stone plinth supports the cruck at ground level.

Tudor/Stuart/William and Mary

A fine house with an outstanding hooded door: the shell hood is resting on carved brackets. Here is elaborate pargetting and square leaded lights. The date above the door of 1692 refers to the refronting of the original house of about 1500.

This is a particularly fine example of pargetting, a medium very common in Essex and East Anglia. The plaster panels are decorated with roses, foliage and garlands. The low post and chain fence is a pretty feature.

Tudor/Jacobean

A double gabled house with an unusual miniature gable in between.

A Hampshire timber framed house with brick infill between the studs, a construction known as brick nogging. The upper storey is jettied and barge boards frame the gables. It is dated early 17th century, according to a date mark, but perhaps its origins are earlier.

Tudor/Georgian

An Elizabethan house with jetty, below which are two 18th century bay windows, and in the roof two dormer windows with hipped roofs.

There is elaborate pargetting along the concealed bressummer and the plaster dolphin was a 17th century inn sign. The plaster has been pitted with a pointed stick, a finish called scumbling.

Medieval/Tudor/Georgian

This terrace without exception consists of timber framed buildings of the 15th and 16th century, all of which were renovated in the late 18th century to give the classical appearance which was fashionable at the time. The fronts date from about 1770, when Georgian door cases and sash windows replaced earlier Tudor features such as small casement or mullioned windows and gables.

Medieval/Tudor/Georgian

A good example of a timber framed house given a Georgian face lift. Rafters in the roof space confirm that the front was gabled, similar to the gable at the back of the house. The roof structure includes a crown post, but internally the house is completely Georgian (about 1780). The timber framing of this building was concealed by plaster. Inside, beams were boxed in, casements were replaced by double hung sash windows. A fine classical door case with recessed door and fanlight was skilfully designed and constructed and situated in the centre of the building so that there is a clear view from one end of the house to the other. Yet at the back of the house many of the earlier features remain.

Medieval/Tudor/Georgian

A handsome house. Late Medieval frame and then a superb renovation in the 18th century. A handsome door with columns of the Doric Order and a parapet with modillioned cornice. The upper storey still projects as it did originally. Often the 18th century facelifters filled in the area below the projecting first floor to give a straight elevation and a larger ground floor area.

Note the skilful design and construction of the bay windows and the pediment with its modillions.

This house originally had three gables to the front, evidenced by the cut off rafters in the roof space.

The three-storey house to the left is brick fronted, probably early 19th century work. Concealed behind that facade is a timber framed building of the 16th century.

Tudor/Georgian

A hall house. An interesting Essex farmhouse of traditional appearance. "Built in the 15th century on the usual Medieval plan with Great Hall in the middle, a Buttery wing on the south and a Solar wing on the north. In the 16th century the Hall was divided into two storeys and a chimney stack inserted". This report from the Royal Commission on Historical Monuments (1916) provides a good indication of how houses have been altered over the centuries.

This house is in Radwinter, Essex. William Harrison (1535–1593), Rector there in 1559, in his *Description of England*, referring to things "marvellously erected in England", speaks of "the multitudes of chimneys lately erected".

Prior to the 16th century, only the larger houses enjoyed the convenience of a chimney. Others burned logs on an open fire, the smoke escaping through a vent or louvre in the roof – hence many ancient houses show smoke stains on the roof timbers.

Tudor/Georgian

A terraced cottage, circa 1500, with jettied storey – it may be earlier.

Renovations have included replastering. The old plaster has been removed and new laths nailed to the timber studs. This method of construction was traditional for hundreds of years. Then expanding metal laths were used. Listed building consent now encourages repairs to be carried out in the traditional way, hence these laths await the plaster – a lime plaster instead of recently used cement!

This cottage has sash windows and a panelled door of the late 18th or early 19th century.

Tudor/Georgian/Regency

A large T-shaped farmhouse. A Tudor or medieval house, much altered in the late 18th or early 19th century. Here is an important chimney stack with diagonal shafts of the late 16th century, delightful Gothick style casements of the early 19th century, a Regency feature perhaps, and prominent bay window of the mid 19th century.

The corner window has been skilfully constructed and is certainly unusual.

Tudor/Regency

What a lovely house? Probably of Tudor origin and then altered at the end of the 18th century, when the delightful Gothick windows replaced those of an earlier period. Note the slender glazing bars.

Five gables with narrow barge boards, carved bressummers at eaves level, a fine door case where classical columns support the porch roof with frieze and cornice.

Was the original house the triple gabled centre?

Tudor/Victorian

A Tudor house known to have been built in 1500 or thereabouts. Externally the lower storey is of brick and the first floor timbers are hidden behind plaster, but the quality of the building is apparent from the interior where the studs and the beautiful moulded ceiling beams are revealed.

In the elevation shown, the upper storey projects and rests on curved and moulded angle brackets.

The two chimney stacks have stepped bases above the eaves and shafts of different design with diagonal pilasters.

Medieval/Victorian

Here is a mid 19th century restoration, creating an elegant facade with tall Doric pilasters framing the door case and sash windows. The side elevation shows the irregularities of a timber framed house, and inside massive timbers suggest Medieval origin.

Tudor/Victorian

A Tudor house, much restored at the end of the 19th century: according to a date mark on a rainwater hopper, in 1882.

The front elevation was given a facelift so that the exposed beams are for ornamentation only and conceal the original oak frame. Notice how straight and equally spread are these studs, unlike the irregular features of an original frame.

The area below the original upper jettied storey has been enclosed with brickwork, enlarging the ground floor rooms.

Inside many of the original Tudor features remain and are exposed. The chimneys date from the restoration.

The stag displayed on each gable is the mark of the estate to which the house once belonged.

Tudor/Georgian

An important timber frame hall house. An extravagant late 18th century restoration with castellated parapet between the gables, much moulding, moulded lintels above the windows and to the gables. A fine doorcase with classical columns.

Attached to the house is the brick kiln for malting barley, an 18th century addition.

Tudor/Georgian

A timber framed house, reconstructed in the middle of the 18th century with many classical features. A symmetrical facade with a fine door case with a canopy supported on Doric columns, and sash windows. The return frontage with its exposed timber on the left shows the original structure. Note the quality of the sash windows, the cornice and the traditional patterned plaster.

Tudor/Georgian

A period terrace. The cottage with a bow appears to be an infill. On the right a Georgian front on an earlier timber framed house.

The windows on the first floor are double hung sash with side lights, and on the ground floor Venetian windows, a triple opening where the centre is arched and the sides have lintels, sometimes called a Palladian window after the work of the great Italian architect Andrea Palladio (1508–1580). This front is circa 1770.

The great chimneys were perhaps added to the original Tudor house. William Harrison, in *The Description of England* written in 1587, refers to "the multitude of chimneys lately erected".

Tudor/Georgian

A timber framed house with a late 18th century front which retained the projecting upper storey. Pretty bays fit below the jetty and first floor sash windows have replaced earlier casements, although casements have been retained in the return frontage. A hipped mellow tiled roof.

The projecting storey running the entire length of the building is called a "long jetty", a form dating the building to the 16th century.

A lovely house. The designers of the day were surely artists.

(If Stansted Airport is extended this house will be demolished.)

Tudor/Victorian/Modern

A small Essex farmhouse, much restored. Behind this modern facade with its standard casement windows is concealed an old timber framed building of the 16th century, perhaps earlier.

The projecting gabled upper storey suggests an earlier building, though there are few other revealing external features. The chimney is recent and belies the great inglenook within.

Elizabethan/Jacobean

Cheshire cottages with timber framing in the form of panels called square framing (unlike the timber framing of Eastern England where usually the timbers are vertical with horizontal timbers at the level of each storey). Here the panels are filled with brick, a common form of construction which contrasts again with the east of England's traditional wattle and daub.

Large dormers extend the accommodation in the roof space. Elizabethan/Jacobean?

Late Stuart/William and Mary

A striking city house, late Stuart with so many lovely features of that period.

A hipped roof with hipped roof dormers, a modillioned cornice, double hung sash windows, and a string course of slightly protruding brickwork between the storeys.

Its date is 1682.

Late 17th Century/William and Mary

A late 17th century house, William and Mary.

These mullioned and transomed double light windows suggest the 17th century. A simple doorcase with delightful fanlight. Above the first floor windows is a modillioned cornice. Presumably the roof was originally at this level.

The top floor, with its many gables, was probably added in the late 18th century. Note the balustrade to the flat roof and the plain barge boards with finials.

Late Stuart

A late 17th century stable building, now converted into three dwellings. Built in 1684, it is a good example of the period; often stables displayed important architectural features and were built to grand designs.

Here to the ground floor is a central arch with square headed windows and door cases of the period on each side. The ground floor windows have a central mullion and a transom. The first floor windows have segmental arches. A parapet partly conceals the roof and the building is surmounted by a fine cupola with clock, a feature of stables of the period, which were often designed and built comparable in quality to the principal house.

Queen Anne

This house dates from 1715, for the year is recorded on the rainwater heads. Here is an elegant facade of mellow brick. The windows, curved or segmental headed, have keystones. An impressive front door with a segmental pediment, and the door pilasters are based on the Corinthian Order. Peeping above the parapet are dormers which have alternate segmental and triangular pediments, a common device of the period. The dormers have casements, not sashes as in the other windows, a fashion of the period.

As a matter of interest, this particular house had Victorian bay windows inserted on the ground floor in the 19th century. In 1951 the front was rebuilt and the bays were replaced with sashes and the building restored to original form.

The pretty railings to the front door are probably contemporary with the house.

The railings along the frontage are Victorian.

Queen Anne

An early 18th century town house, perhaps Queen Anne, circa 1700–1720.

A central projection of the facade is surmounted with a pediment, and on either side a cornice and parapet conceal the hipped roof. Double hung sash windows and the door are framed in stone. The door has rusticated pilasters, an arch and a segmented pediment with cornice.

Tudor/Queen Anne

A striking town house. A good 18th century brick front on an earlier timber framed house. Once an inn.

A pretty door case with canopy, steps and railings. The front is about 1720. The many sash windows probably once had smaller panes. A modillioned cornice and hipped roof. A good example of its period.

Note the wrought iron support from which once hung the pub sign.

Tudor/Georgian

This group of houses are of Tudor origin, refronted in the mid 18th century. The sash windows on either side of the left-hand door show 16 panes and are more recent than the remainder where the smaller windows with thicker glazing bars suggest the 18th century; the right hand door and the dormer windows date from the mid 18th century. The roof is hipped at the left hand end.

Elizabethan/Georgian

A bold town house with steps and railings.

This is an 18th century front on a much earlier timber framed house. The bays were added in the 19th century.

The timber frame was completely enclosed with brick, stucco rendered and the roof was raised and covered in slate, which was becoming a popular roofing material from the late 18th century.

This doorway would have been constructed by a local craftsman from a design in a published pattern book. A delightful classical door case with Tuscan columns, and a rusticated frame to the arched door with its lovely fanlight . The columns support a frieze and pediment.

Tudor/Georgian

The massive chimney with its diagonal shafts gives a good indication that this house belongs to an earlier period than the front suggests. The many features of this early Georgian facade include the rusticated brick pilasters to the door, the panelled parapet with a projecting cornice, and brick quoins.

This timber framed house of the Tudor period was refronted in brick in the 18th century, circa 1750, at a time when similar restorations and alterations were taking place all over England.

The door case and its surround alone are a work of art, with gently rising steps and the wrought iron hand rails.

Early Georgian

An 18th century terrace of cottages with brick elevations, the gable end supported by a brick buttress and the S iron brackets secure the metal ties running from front to back, installed to prevent further movement in the brickwork. The roof is thatched with combed wheat reed, common in the West Country.

The brickwork is a good example of Flemish bond, where headers and stretchers alternate.

Georgian

A Cheshire cottage built from local stone (and pretty massive stones too) with a stone tiled roof and simple casement windows. A low pitched roof with shallow verge parapet at each gable. Probably 18th century.

The casement windows would presumably once have been glazed with small panes with glazing bars.

Georgian

These cottages at Milton Abbas were built in 1773–1779 as part of the new model village, and developed in a location away from and out of sight of the great country house, Milton Abbey. The cottages, equally spaced, stand back from the village street with its wide grass verges.

Construction is of oak frame and cob under a combed wheat reed roof. The front door originally led to two separate cottages, now converted to one.

The plain doors and casement windows are good examples of the simplicity of design used in cottages of the period.

Georgian

An elegant town house.

An important house of the mid 18th century with shell hooded doorcase, flat arched windows with keystones and stone sills. Contemporary railings to the front on low or dwarf walls. A shallow pitched roof screened by a low parapet with cornice.

The fine old wall to the left is capped with coping stones.

Tudor/Georgian

A West Country stone fronted house, the earlier timber frame is clearly exposed in the side elevation, with its square framing. Well designed sash windows of the late 18th century at one end and earlier casement windows at the other. A Georgian stone front on an earlier timber framed house, or quite possibly it may have been built at the same time.

Georgian

Perfect symmetry – an early Georgian house.

Here are double hung sash windows with brick arches and a string course between ground and first floor. The central doorcase and the panelled door are an important feature. This classical doorcase has a broken entablature, triangular pediment and fanlight, and Doric pilasters of about 1750. The pretty cornice conceals the guttering and above are flat roofed dormers. Note the verge parapets.

Plain, railings complete the picture.

Early Georgian

A fine early 18th century house. While of earlier origin, what we see is of 1731 – early Georgian. A broad doorway flanked by classical columns and crowned by a segmental pediment; an original panelled door.

Sash windows, with segmental arches with keystones, to the ground floor and square headed windows to the first floor.

A wide soffit and flat roof dormers with sashes. A good example of its period.

Georgian

Here is a spectacular residence which is completely balanced. A doorcase with a broken entablature and triangular pediment framing the fanlight. The windows set flush with the external face of the wall which marked the Queen Anne and early Georgian period. London Building Acts 1708 required recessed windows, although this style took some time to reach the provinces.

The various features of this house, particularly the doorcase, suggest a date of around 1740.

Georgian

An 18th century house with an abundance of windows in the style of the Italian architect Palladio and commonly known as Venetian or Palladian windows.

The windows are of three lights with a central semi-circular arch and flat arched side windows. Here the door is of similar design.

Note the lovely delicate glazing bars. The bay windows are possibly 19th century additions.

Georgian

A row of brick and tile 18th century cottages with doors and casement windows of the period. The flat roof dormers are probably contemporary.

A brick and flint cottage. The principal material is flint, with brick courses between the ground and first floor windows, patterned brick and flint work. A thatched roof of combed wheat straw and eyebrow window. A jolly design. In many areas flint is available from quarries and fields.

Georgian

A delightful and very good example of the use of clunch – chalk blocks – in this late 18th century terrace. At the corner are brick quoins. The roof is hipped. Large sash windows to the ground floor and casements above. Charming door hoods.

Georgian

A late 18th century front. Note the panelling between the bays, an interesting feature.

A handsome door case, flat roofed dormers above the parapet which has a moulded cornice below. Here is a tiled mansard roof.

A fine carriageway with semi-circular arch with a keystone, which gave access for wagons to the maltings at the back.

An important building of its period, doubtless the house of a wealthy maltster.

Georgian

Tudor

Compare the austere yet elegant features of the mid 18th century stone house (top) with the elaborate detail of the late 16th century brick house. Here is a good example of brickwork of its period, and original stone windows and gables. The ornamental gables are surmounted with finials. The rainwater pipes are original and a feature.

Here is a clock turret with cupola and bell. The gabled ends are surmounted with finials.

The delightful stone house enjoys many dignified features of the 18th century, including a panelled front door framed by Tuscan columns supporting frieze and pediment; lovely bow windows with sashes and side lights, and to the first floor, double hung sashes with side lights.

Dividing the two storeys is the stone string course and all is crowned with the modillioned cornice with central pediment and parapet. The architect would have been pleased with his work, which he finished off with painted railings.

Georgian

A bold building with a mansard roof, and a parapet with pierced balustrades below which is a prominent cornice.

The paladian moulding in the form of a sealed window is an important feature, as is the scrolled pediment above the door.

Georgian

A Georgian cottage, small but with important features including a door case with classical pilasters and shallow canopy and sash windows.

The roof is covered in pantiles and the principal building material is flint – a local stone and a cheap material with an abundance of flints in every field. The windows have brick surrounds.

Georgian

A small town villa of the late 18th century. Doric columns support the frieze and modillioned cornice of the porch. Recessed sash windows, a deep soffit and rendered elevations create a plain and elegant building.

Georgian

This house, originally perfectly symmetrical with its centre door, appears to have been extended by the addition of wings on either side. That to the right has a gabled roof and that to the left a hipped roof, while the principal roof is hipped. The wings are of different length and one has a bricked up window – presumably to reduce the window tax liability. The double hung sash windows have straight stone arches with keystones.

A handsome door case with cornice, frieze and pediment supported on pilasters based on the Corinthian order. The railings appear to be modern.

Mid 18th century with later additions.

Georgian/Regency

An 18th century farmhouse with an early 19th century wing.

The principal house may have earlier origins, but all the features suggest the mid to late 18th century – sash windows, a classical door case, dormers with hipped roofs. The wing of perhaps the 1820s is of brick (the original being lath and plaster). Here are deep sash windows in a recessed arch extending into the gable or pediment. The fine old flint garden wall appears to be earlier than the house.

Late Georgian

A pretty house with its important door case with bay windows with sashes on either side, and on the first floor casement windows with an elaborate pattern of glazing bars.

A late 18th century facade.

The roof is covered in clay tiles except for the eaves where there are two courses of local stone tiles.

Late Georgian

A striking house. Such elegant features; sash windows with sidelights and fanlights, and on the ground floor a delightful panelled door with side lights and panels and matching fanlights. A hipped roof.

Steps and railings lead to the front door and the wall along the pavement adds to the completeness of the design. Late 18th century.

Late Georgian

A late 18th or early 19th century cottage of brick and flint. The flints were doubtless collected from the nearby fields.

The wide brickwork band between the first floor windows is patterned with flintwork shapes, and other brick courses frame flintwork panels.

Above the ground floor windows are segmental arches.

The first floor windows are iron casements.

The builder obviously enjoyed his work and created a "jolly" building.

Late Georgian/Regency

A pretty, restrained terrace with all the good features of the period. Stuccoed elevations, slate roof. Elegant sash windows and arched doorways. A broken cornice provides shallow hoods for the doorways. Contemporary railings. A charming terrace.

Regency

An elegant villa, early 19th century, with pretty doorcase with fanlight with glazing bars, and here is a bay window supported on brackets.

The first floor Gothick style windows, with contemporary canopy above, are a special feature of the house.

Please note that Regency Gothick is usually spelt with a "k".

Regency/Early Victorian

An early 19th century terrace.

    The grandeur and boldness of the 18th century is evident.

    Notice the gable in the form of a pediment, the two storey bow windows and the bold front door with Doric columns.

    The front elevation is stucco rendered. The side reveals the brick construction where huge sash windows suggest an important first floor room.

    An elegant Regency terraced house suited to a country town.

Regency

A charming Regency villa completely unchanged since it was built.

Delightful windows with ogee arches, a portico, again with ogee arches, and a balcony above with wrought iron balustrade. Pretty, shaped, modillions enhance the soffit.

An artist's design! The ogee design was a feature of 18th century Gothick revival. Nothing disturbs the balance of this beautiful facade.

Regency

A house of distinction. An important town house with an elegant, completely balanced, facade in stucco. An imposing portico with Doric columns supporting the entablature (of cornice and frieze), an arched door and graceful bows on either side, and five sash windows to each of the upper two floors.

There is a wide soffit and the low pitch conceals the roof.

This is a most striking building of immense quality in both design and construction. About 1820.

Regency

A pretty city house of the Regency period, Here are elaborate barge boards on twin gables with finials, a wide soffit, elegant sash windows and a pediment door case, and, of course, the Tudor style chimneys are an important feature.

Regency

Here is a pretty terrace, and each pair of cottages is stepped. Built of brick, with tiled roofs, this neat and tidy terrace has pretty door cases with pilasters and shallow canopies.

The windows are iron casements with brick arches above. Iron work came into its own in the second half of the 18th century. The date mark is 1810.

Late Georgian/Regency

A town house with a plastered front of the late 18th century. Here is complete symmetry. The door case with fanlight is set in a recessed arch above which is a keystone.

On either side of the door are bay windows to both ground and first floor. Note the thin and elegant glazing bars of this period (upstairs). Earlier glazing bars were much thicker.

At the angles of the facade are large quoins. The parapet is pierced with balustrades above the modillioned cornice.

Regency

A delightful estate cottage of local flint and brick. A skilful and enchanting design – completely balanced.

A central door in a projecting bow with Gothick arched windows and the extended roof supported on timber studs to provide a verandah.

It is difficult to adequately describe such a beautiful feature of the English countryside.

Regency

A jolly house in the Gothick style of the Regency period – early 19th century. It has a castelated parapet, stone windows with mullions and transoms, and above each a moulded lintel. A pointed arch to the front porch and elaborate barge boards.

An elaborate chimney and gabled parapets complete the picture.

Late Regency/Early Victorian

This delightful lodge or gate house is built of brick with stone dressings. A two storey stone bow window is surmounted by a pierced parapet, and there are Tudor style chimneys and a tower or cupola. A stone doorcase. Built in 1834.

The bold ornamental gate posts adjacent to the lodge are of the same date.

Georgian/Victorian

A good town house, a date mark of 1848 confirms its period. A good example of a plain, late Georgian, house built at a time when designs were becoming more flamboyant and high Victorian Gothick was about to replace the simplicity of the classical design. Deep sash windows, a plain front door with fanlight but no supporting door case. A wide soffit and shallow pitched roof.

Mid Victorian

A fine example of a 19th century farm house, perhaps replacing an earlier building.

It is of cobblestone construction, a material often used in eastern England. This is an impressive house with large sash windows with side lights to both first and ground floors. Verge parapets frame the pantiled roof and a single storey extension with lean-to appears to be original. The quality of this house suggests a prosperous farming period. Note the pretty iron entrance gate with its hoop and bars.

Mid Victorian

Two houses, miles apart, but perhaps of similar date, circa 1850. They certainly share common Italianate features, which are modillions to the soffit and semi-circular arched windows, and a shallow pitched roof.

That on the left has rusticated quoins and an important portico with Corinthian columns.

Immense thought went into their designs. Note how the front door canopy joins and supports the first floor window in the top house.

Georgian/Victorian

Here is a striking residence. Built by a gentleman for his own occupation. It enjoys all the features of the Georgian period but in fact belongs to the early Victorians – about 1850.

What makes it immediately impressive is its four square angle turrets at each corner, surmounted by lead covered hoods shaped rather like helmets. Here are castelated parapets, a cornice below and a string course between the ground and first floor windows (deep sash windows with delicate glazing bars). The exterior suggests fine proportioned accommodation inside.

The hoods to the turrets are similar to that of the lodge (page 94); perhaps the same architect?

Victorian

A pretty village house, part of a great estate whose noble owner left his mark and date of construction – 1860. The bricks were produced on the estate.

Unusual first floor windows surmounted by a pointed arch with pierced barge boards and, to the ground floor, bay windows with hipped roofs and casements. A string course at first floor level. The massive chimney suggests a fireplace in every room.

Victorian

A good example of High Victorian Gothick architecture – a striking building. Built of red brick with stone dressings, sweeping roof crowned with open crestwork to the ridge. There are bold square bay and lancet windows, with pointed arches, patterned brickwork, grand chimneys and many Victorian Gothick features. The irregular shape of the house is the outside expression of convenient planning inside. 1864.

Victorian

A pair of late Victorian estate cottages with an abundance of features. An impressive chimney, crested ridge tiles, barge boards, ornamental brick arches above the lintels of the pretty casement windows, string courses of brickwork, steep pitched roof porches and much else.

Victorian

A late 19th century house, perhaps originally a school room.
  A fine example of ornamental Dutch gables and fancy Tudor style chimneys.
  Pointed ridge tiles to a slate roof.

Victorian

A pair of farm cottages of 1874, built of flint with brick dressings under a slated roof with hipped gables.

Brick arches to the windows and brick quoins at the angle of the walls.

If the porches are later they are skilfully designed to match the original.

Victorian

A bold estate house, late Victorian and typical of that period. The house follows the design of a Tudor Hall House, with gabled wings, barge boards and finials and crest work to the roof ridge. Large and important chimneys designed as an impressive feature.

Victorian

A mid 19th century cottage with weather boarded elevations. Farm barns and other buildings also commonly used weather board, which was not so frequently used for domestic buildings although there are many examples.

Late Victorian

A late 19th century farmhouse – a pleasing creation with multi-curved Dutch gables and matching parapets to the windows. A bold porch with steep pitched roof. A central chimney with four shafts and a date mark and arms to the base. Here the four downpipes are designed as a feature dividing the bays.

Victorian

Late 19th century terraced houses, thousands of which were built in towns all over the country at the end of the 19th century. Such ubiquitous terraces played an important part in the creation of the suburbs. Bay windows and integral porches were a common design feature. The multi-potted chimneys indicate a fireplace in every room; the only form of heating at that time. They are generally very well built and will last indefinitely.

Late Victorian

A pair of semi-detached houses faced in stone and tile hung to the upper storey, with fishscale tiles to the gable.

A very attractive design with a variety of windows and a pleasing turret with weather vane. Here is a steep pitched slate roof with open crestwork to the ridge and a finial to the front gable. About 1890.

20th Century Copy of Tudor Style

A "black and white" house (actually a bank), typical of the district.

Two oversailing or projecting storeys. The timber work is similar to many buildings in the Cheshire area. Barge boards with finials enhance the building. The straight beams denote the fact that the building is mock tudor.

It became fashionable to treat exposed timbers with tar or pitch in the early 19th century, creating the colour contrast between the timbers and the plaster. This building, a replica, was built in 1900.

Victorian/Edwardian

The top picture shows a terrace built just before the turn of the 19th century and a detached house built just after. The latter is marginally plainer. No modillions, no doorcase nor canopy.

Edwardian

A row of detached houses of the early 20th century. Here are sash windows with large panes, and no glazing bars. A plainer design than the Victorian houses which preceded them. There are few frills except perhaps the finials prominent on the roofs.

Edwardian

An early 20th century town house. Bay windows with large panes were a feature of this period. Here is some mock Tudor timbering to the gable and a prominent and distinguished dormer window. The chimney plays a prominent role in this design.

Edwardian

Houses of the Edwardian period. Semi-detached, they copy the contours of a Tudor hall house.

A pretty verandah style double porch with elaborate wrought iron support sits between the wings and is an imposing feature. The pitched and hipped roofed bay windows are typical of the period.

Other features include the modillioned cornice above the central windows, the string courses of light brickwork, bargeboards and open crest work to the roof ridges.

Early 19th Century and Modern

A recently constructed terrace of new houses, built on the site of a former 18th century brewery. The classical features are in keeping with adjacent properties.

The return frontage shows the side elevation of the original brewery building, now restored and forming part of this residential complex.

1920s and 1930s

An art deco design with bow to the wing and metal windows. These metal casements were very popular between the Wars.

A distinguished house of the period. A plain design save for the Dutch gable.

1920s and 1930s

Two houses emulating the Tudor period, principally by exposing fake studs and beams.

Modern

Despite their appearance these two houses are only 10 years old as a close inspection would reveal. They are designed and built in the style of the early and mid nineteenth century. Excellent workmanship has produced a delightful Regency-style porch (house below) and a fine doorcase (house above) as well as handmade sash windows.

Sash windows (sliding up and down one in front of the other) were invented in Holland towards the end of the 17th century and arrived in England about 1685. Early examples have thick glazing bars, which became progressively more slender during the 18th century.

Modern

A carefully designed terrace in the vernacular style. Double hung sash windows and traditional door cases with canopies. Note on the left of the picture the impressive bay of a Regency house.

N.B. Vernacular: "of the jargon or idiom of a particular group: native, local, endemic, especially of architecture or general style of building."

17th Century

Modern

A carefully designed row of cottages, recently built in the vernacular style and constructed in stone. Casement windows to the ground floor and dormers above. Steeped roof canopies to the doors. Dry stone walls along the boundary. Very pretty.

Compare this with a 17th century group of similar construction (top).

Modern (1930s)

Semi-detached houses. These are typical of thousands in similar style which mushroomed all over England between the wars, 1920–1940. Many had bay windows with hanging tiles, and arched porches leading to the front door.

Their elevations were brick or, as in this picture, they were often pebble dash rendered, or roughcast.

Hipped roofs were often a feature.

In some cases there was no provision for a car.

Modern (1960s)

In the 1950s and 1960s, smaller houses were often plainer and therefore less expensive to build. It was fairly standard for the houses to be built to include a garage.

Barns

The top picture shows a conversion of a massive timber framed barn. The windows have been constructed so as to leave the studs undisturbed.

Traditional farm homestead buildings, with barns, cart lodges, cowsheds and stables near to the farmhouse, often dating back to the 14th century, have become obsolete for modern farming practices.

As a result "Barn Conversions" have become a feature of the countryside. Some have been skilfully designed, restored and converted, retaining historic characteristics, but others are less successful.

Here, in the bottom picture, a great timber framed barn with midstrey (that is the carriage arch projecting in the middle) has become a house. The roof is thatched in Norfolk reed, weatherboard to the ground storey has been retained and the upper storey is rendered.

Barns/Georgian

A massive 18th century barn of clunch, with a pantiled roof – materials common to this area, East Anglia. Perhaps awaiting conversion to houses.

The quality of the construction is striking; built on a great estate to a standard as high as that for domestic buildings.

Stables Early 19th Century

19th century stables and chaise houses. Brick and pantile roof, arched doorways. A fortunate survival, but for how long? This might be viewed as ripe for conversion to cottages.

London

"When a man is tired of London, he is tired of life; for there is in London all that life can afford."

So spake Dr. Johnson in 1777.

Architecturally, at least, that is probably true today. It would be impossible to describe the vast variety of buildings; houses, flats, and apartments that make up the amazing place that is London.

Here is a small selection of mainly 18th and 19th century dwellings.

Involved were some of the great architects of those years whose contributions to the quality of London is indisputable. Their enormous skill in design, engineering and art is to be marvelled at. Among others, we have to thank:

Robert Adam (1728–1792), John Nash (1752–1835), Sir John Soane (1753–1837), Sir Charles Barry (1795–1860), Edward Blore (1787–1879), Richard Boyle, Earl of Burlington (1694–1753), Sir William Chambers (1723–1796), Henry Flitcroft (1697–1769), James Gibbs (1682–1754), Nicholas Hawksmoor (1661–1736), Henry Holland (1745–1806), William Kent (1685–1748), James Paine (1717–1789), Sir John Vanbrugh (1664–1726), Lewis Vulliamy (1791–1871), Sir Christopher Wren (1632–1726), James Wyatt (1746–1814), and many more.

Readers wishing to know more of those architects and their works should refer to the amazing reference work *A Biographical Dictionary of British Architects 1600–1840* by Howard Colvin.

London

These houses date from 1720 or thereabouts: Queen Anne and early Georgian. Many of London's residential streets appeared like this in the 18th century.

The houses of this period are of mainly simple designs, rectangular in shape and plan with sash windows and distinguished front doors and doorcases with hoods and fanlights.

The London Building Acts of 1707 and 1709 required window frames to be recessed not less than four inches from the outside wall. Perhaps it took some time before developers complied with the new legislation.

London

A terrace of the late 18th and early 19th century. Once residential, the ground floor became shops, probably early in the 20th century. The deep set sashes have stone surrounds and above the modillioned cornice a balustrade forms a parapet.

London

This pretty terrace, set back from the street, is of the Queen Anne period.

Note that the window frames are flush with the walls. London Building Acts of 1707 and 1709, both concerned with the threat of fire, required window frames to be recessed by not less than four inches from the outside wall; wooden eaves cornices were abolished and the front wall was to be carried up above the roof as a parapet.

The ground floor of this terrace has later features and was probably altered in the early 19th century.

London

An imposing terrace in a style common to many English towns. Late Georgian, it has grand doorways with rusticated pilasters and the fanlights are typical of the period – about 1775. Here are railings at street level, a pretty railed balcony with canopy to the first floor, and cornice above.

London

A London residence of the late 18th century. The end of the terrace reveals the brickwork construction, faced in stucco. Good features of the period include a fine doorcase with prominent fanlight. Above the centre first floor window is a pediment, a bold cornice above the third floor, which is returned along the side elevation. A pretty wrought iron balcony and glazing bars to the side windows. At the angle of the building are the dressed stones called quoins (pronounced "coins").

London

An elegant, plain, three storey, terrace. Well, not too plain but reserved, circa 1830.

Recessed front doors, unframed sash windows; the architect allowed himself some ornamentation to the first floor with small balconies and hoods to the windows supported on brackets. The parapet conceals the roof and along the pavement are contemporary railings.

London

A London Terrace. Early 19th century elegance. Perfection in design; on the ground floor an imposing entrance with Ionic columns supporting a balcony; first floor arched windows with keystones; a pretty cornice above the third floor windows. Railings of the period complete the picture.

London

This terrace, stucco faced, expresses order and elegance.

Here is a verandah where Doric or Tuscan columns support a balcony with frieze, cornice and balustrade.

Pediments surmount the first floor windows, while above those on the second floor, are moulded lintels.

The elaborate cornice of garlands is an important feature and quoins decorate the corner of the facade.

An important London Terrace built between 1826 and 1855. Originally constructed as private houses, after the Second World War many were converted to luxury flats and apartments and, in some cases, offices.

Fine residential squares were developed in the early part of the 19th century for the great land owning families whose estates included freehold land appropriate for London's rapid expansion.

London

Mews. Originally the yard and stabling for important houses, later often converted to garages. More recently perhaps from the middle of the last century, stables and garages have been converted to houses and provide appealing 18th and 19th century dwellings, often in quiet and private locations.

London

A Regency Terrace with pretty verandah, where floor to ceiling glazed doors open on to small balconies. This stucco faced building is typical of its period.

London

Two nearby terraces of the early 19th century.

In the one, conventional square headed sash windows to the ground floor and lancet windows to the first floor, and in the other, quite by chance, just the reverse.

Houses of this style present an attractive street scene with simple lines and lack of over ornamentation.

London

A delightful late Victorian terrace about 1875. Undoubtedly the work of a skilful architect. Here are bays, balconies, balustrades, cornices, rusticated work and much more. A most impressive, bold and confident design.

London

A Victorian terrace, thousands of which mushroomed in the suburbs from about 1870.
They were substantially built, and often included bay windows with slate hipped roofs, and entrance porches. The stonework was embellished. Inside, spacious accommodation included high ceilinged rooms of good proportions.

London

An important Town House incorporating so many features as to make it appear overcrowded. The pediment is supported by twin Ionic columns and there is much rusticated work to the ground floor (that is stone blocks or bricks separated from each other by deep joints). The first floor bays are surmounted by parapets, and there is an attractive parapet pierced with balustrades to conceal the pitched roof and dormers. Probably early 20th century.

London

A London Mansion Block. Many in similar styles were built at the end of the 19th century and provided spacious flats with good, well proportioned, light rooms with large windows. Balconies compensated for lack of garden.

Here, arched porticoes support balconies with stone balustrades, while elsewhere the balustrades are of iron.

London

Here is a massive residential block of the late 19th century.

An immense amount of planning and design went into that. Built of brick with stone dressing, there are bays and balconies, sash windows with a variety of glazing bars, classical columns, a corner turret, dormers and elaborate chimneys, not to mention the grand entrance of wide steps with stone balustrades.

London

A block of flats in the streamlined form of art deco, a popular style in the 1920s and 1930s, with little ornamentation. Here are simple lines and both sash and casement windows.

Almshouses

Almshouses were built in most towns and villages from the 17th century and were endowed by Trusts, Charities, and individual benefactors to provide lodging for the poor.
Here are two examples.

Above, village almshouse of the Regency period with elaborate barge boards, a Gothick window to the first floor, with moulded arched lintel and a casement window to the ground floor with mullions and transom and moulded lintel.

Below, more important almshouses with a central chapel. Built of brick and slate and completed in 1834, they present an imposing group. The centre gabled chapel has an arched doorway with lancet windows on either side. Pinnacles surmount the gable, and here are parapets and buttresses.

Local Authority Houses

After the First World War legislation enabled local councils to build houses within their districts, to be let at reasonable rents to help meet urgent post war housing needs and to replace substandard housing.

What were called "Council Houses" mushroomed all over England. They are recognisable for their style and uniformity. More recent legislation allowed the sale of these houses to their tenants at discounted rates. Consequently, the uniformity is often breached as individual owners do their own thing; erect porches, alter the windows and colour for example.

A Miscellany of English Architectural Features

A small sample of English street furniture and other features which give one a feel for good design; work to appreciate.

An 18th century style cupola. An elegant design with simple columns, open arches, a modillioned cornice, all surmounted by a beautiful scrolled weather vane.

A substantial late Victorian water tower, designed and built with flair and style despite its purpose. Elaborate brickwork and an elegant cupola with weather vane.

An original 18th century bow window with contemporary panelled door. A narrow canopy above the frieze is supported by four modillions and a single bracket supports the window. There was doubtless another door at the right hand end, one door would have led to the shop the other to the house.

A boot scraper. Before metalled roads and pavements, muddy footwear required scraping clean and scrapers were a design feature of many front doors. An abundance of 18th and 19th century examples survive.

The gateway to a great country house. Built in 1778, the work of the celebrated architect James Wyatt. Here are symmetrical lodges on either side of the arch (a semi-circular arch) with Doric columns surmounted by frieze, cornice and pediment. The lodges have double hung sash windows. Many door cases of the period in town and country resemble this arch. Compare with the example on p 61.

A street light of the early 20th century. A ubiquitous piece of street furniture (an iron standard and a copper lantern) was to be found in most English streets. Some still survive, converted from gas to electricity. An exquisite design.

Examples of wrought iron work date from the early centuries BC, followed by medieval railings and door furniture, decorated wrought iron scrollwork and Victorian church screens.

Of particular note are the garden railings of the 18th and 19th centuries, which adorned many town and country houses. These are indeed an important feature of the English country house and usually worthy of note.

Sadly, however, railings were removed during the last War, to be used for ammunition production, but many fine examples remain and should be cherished.

Here, among others, are railings, and wrought iron gates with arch and lamp, contemporary with the church to which they lead, circa 1780.

These are 19th century railings; simple, unobtrusive.

Dry stone walling. A beautiful feature of the countryside. To be found where there is local stone – in the north and Wales for example.

A white paling fence; a rural feature of great appeal.

A flint and brick wall with pitched coping. Walls are a great feature of many properties.

Wrought iron work. Important estate gates leading to a great country house. Brick and stone columns surmounted by eagles and urns.

An early Dovecote. Once apparently a gate house with dovecote above. The unusual crossed timbers suggest an early date, perhaps 14th century.

A classical three arched stone bridge with cornice and parapet pierced with balustrade (circa 1760).

A traditional five bar gate; natural material; functional and fitting.

A timber sign or finger post. Designed for clarity and succeeding. A delightful feature at a country crossroads.

The once ubiquitous telephone box. A superb creation by the architect Sir Giles Gilbert Scott. This design was modelled on Sir John Soame's monument in the burial ground of St Giles in the Fields, St Pancras.

Useful Adjectives for Estate Agents in Describing Houses

Admired
Ancestral
Attractive
Beautiful
Capital
Character
Charismatic
Charming
Classic
Classical
Comfortable
Commanding
Commodious
Compact
Complete
Contemporary
Delightful
Derelict
Desirable
Dignified
Distinguished
Edwardian
Elegant
Elevated
Eminent
Enchanting
Enviable
Established
Excellent
Exceptional
Fascinating
Favoured
Fine
First Rate
Flawless
Formal
Freehold
Georgian

Glorious
Good
Graceful
Gracious
Grand
Gripping
Handsome
Historical
Idyllic
Immaculate
Important
Imposing
Improved
Lonely
Lovely
Luxurious
Magnificent
Majestic
Matchless
Mature
Meticulous
Mews
Modern
Neat
New
Notable
Old Fashioned
Old World
Outstanding
Palatial
Panoramic
Period
Picturesque
Popular
Pretty
Private
Prominent
Quality

Refined
Refurbished
Regency
Remarkable
Remote
Restored
Rich
Richly
Rustic
Secluded
Solid
Sound
Spacious
Special
Stately
Striking
Stunning
Stylish
Substantial
Superb
Superior
Supreme
Sympathetic
Tasteful
Traditional
Ultimate
Unique
Unparalleled
Unspoilt
Useful
Valuable
Versatile
Victorian
Well Designed
Well Maintained
Well Presented
Well Proportioned
Wonderful

General Information Relating to Property Matters

Abstract of Title

A summary or abridgement of the deeds constituting a title to an estate.

Accommodation Land

Land which is to be developed is put to alternative use in the meantime.

Act of God or Force Majeure

Words used to describe an occurrence which man is unable to prevent or anticipate – a great storm for instance.

Acts of Cultivation

The various cultivations carried out on agricultural land, and listed in agricultural valuations for stocktaking or tenant right – ploughing, cultivating, drilling, harrowing, rolling, spraying, etc.

Advowson

The right of presentation to an Ecclesiastical Living. The owner of the advowson, known as the Patron of the Living, has the right to appoint the Rector or Vicar of the Parish.

Afforestation

The establishment of forest plantation on treeless land. (Re-afforestation is planting on felled forests.)

Ancient Lights

Lights, through windows that have remained in the same place and condition, and have been enjoyed for 20 years.

Arboretum

A collection of different tree species grown for exhibition or scientific study.

Areas

One acre equals 4,840 square yards (449.6 m²). Old deeds often refer to acres, roods, rods, poles and perches (rod equals 30¼ square yards). One acre equals four roods; one rood equals forty rods, poles or perches. (One rood equals 0.25 acres).

One hectare (10,000 square metres) equals 2.47106 acres (2½ acres in round figures).

Arts and Crafts Movement

An English late nineteenth century aesthetic movement instigated by William Morris and others. It opposed mass production and sought to revive medieval craftsmanship and materials.

A Biographical Dictionary of British Architects 1600–1840 (Howard Colvin)

An invaluable reference relating to, among other buildings, larger country and town houses.

Boundaries	Hedge and ditch. In the absence of evidence of ownership, stand in a field with the ditch on your side of the hedge, then the boundary of the field in which you are standing reaches only to the edge of that ditch, and both the hedge and ditch belong to the field/garden on the other side. Where there is no ditch, the boundary is assumed to be the centre of the hedge.
Caveat Emptor	"Let the buyer beware."
Conservation Area	"An area of special architectural or historic interest, the character of which it is desirable to preserve or enhance."
Copyhold	A form of tenure held by the Lord of the Manor. The Law of Property Act 1922 converted copyhold to freehold.
Council Tax	From 1993 the Council Tax replaced the Community Charge (or poll tax).
	Each dwelling is placed on a valuation list in one of eight valuation bands ranging from A to H. The list shows only the band to which a house has been allocated, not its actual value.
	Valuation Bands:

A	Up to £40,000
B	Over £40,000 and up to £52,000
C	Over £52,000 and up to £68,000
D	Over £68,000 and up to £88,000
E	Over £88,000 and up to £120,000
F	Over £120,000 and up to £160,000
G	Over £160,000 and up to £320,000
H	Over £320,000

Covenant	A clause in a lease, etc whereby a party binds himself to do or not to do a certain thing.
Dower House	A dwelling reserved for a widow to live in after her landowning husband's decease.
Easement	A right or privilege enjoyed by an owner of land, in respect of his ownership, in or over the land of his neighbour: for example, a right of way.
Estates Gazette Diaries (early ones)	These contain an abundance of useful information – tables, weights and measures, areas and much else.

Freehold	The highest form of land ownership in English Common Law referred to as "an estate in fee simple absolute in possession". A leasehold interest is for a term of years.
Flying Freehold	A separate freehold interest in part of a building, not including the soil. For example, a first floor room in one house may be above the ground floor room of the adjoining house, each in separate ownership.
Ha-Ha	A ditch or sunken fence surrounding a garden forming a boundary which does not interrupt the view either way.
Husbandman	A tenant farmer as distinct from a yeoman.
Kelly's Directories	Frederick Festus Kelly began publication of provincial directories in 1845, and in succeeding years the directory covered most counties recording every town, village and hamlet as well as listing the names of prominent citizens, farmers, businesses, vicars and landowners. It described and dated churches and chapels and other important buildings. There is much of interest to an estate agent in pursuit of local knowledge. Most libraries have some old copies.
Leasehold	Any estate in land which is not freehold. The essence being that it comes to an end after a specified length of time.
Listed Building	A building of special architectural or historic interest which appears on a list compiled and approved by the Secretary of State and which it is forbidden to demolish, alter or extend without consent from the local planning authority. There are three grades of listed buildings as listed below, *viz*:

Grade I   The most important buildings. Palace of Westminster, cathedrals and churches, ancient guildhalls and important stately homes, for example.

Grade II*   Usually older houses in town and country which have special architectural features, early timber framed buildings with crown posts, good Georgian houses.

Grade II   Many houses and buildings with period features, but houses in towns and villages are also listed for their group value.

It is a criminal offence to alter a listed building without consent. When in doubt, refer to the local authority.

Pightle	Any small enclosure of land.
Possessory Title	A right to ownership of property of which the evidence is not conclusive and which is capable of being challenged by one with an allegedly stronger right.
Pre-emption	A right to purchase before another.
Property Misdescription Act 1991	Under this Act it is a criminal offence to make a false or misleading statement and applies to an estate agency business or a property development business. The statements can be written, oral, photographic, or a model. The Act is enforced by local Trading Standards Offices. Check your facts!

Quarter Days

25th March	–	Lady Day
24th June	–	Midsummer Day
29th September	–	Michaelmas Day
25th December	–	Christmas Day

Riparian Owner	Owner of the banks of a river.
Rights of Way	A right to pass over another's land, subject to such conditions as are specified in the grant or by the custom by virtue of which the right exists. A lesser right is always included in a greater right – for example, a bridleway will include a footpath.
Running with the Land	A covenant is said to run with the land when each successive owner of the land is entitled to the benefit of the covenant (or liable to its obligation).

Seasons for Fishing

2nd February–1st November	Salmon (11th February–31st October in Scotland)
2nd March–1st October	Trout

Seasons for Foxhunting   November to April.

Seasons for Shooting

12th August	–	Grouse shooting begins
1st September	–	Partridge shooting begins
1st October	–	Pheasant shooting begins
10th December	–	Grouse shooting ends
1st February	–	Pheasant and partridge shooting ends

Stamp Duty

Land and property:	
£0–£60,000	Nil

£60,001–£250,000	1%
£250,001–£500,000	3%
£500,001 and over	4%
Transfers by gifts	Nil

Title	The legal right to the ownership of property, evidenced by title deeds or by the Land Registry.
Tree Preservation Order (TPO)	An order by the local authority prohibiting the felling or lopping of a tree without consent.
Unadopted road	A road not maintained by the local authority.
Wayleave	A right of way for wires, pipes or pylons maintained by permission of the owner whose land they cross.
Window Tax	A tax first imposed in 1697 and repealed in 1851. It was levied according to the number of windows and openings on houses having more than six windows. Endeavours to avoid the tax led to the blocking up of windows, evidence of which may still be seen in many houses.
Yeoman	Strictly a freeholder cultivating his own land – often used generally for a farmer.

In Praise of Estate Agents

Estate agents are much maligned and, judging from a recent television programme, not without some justification.

However, in every market town there are to be found handsome offices of old established firms of auctioneers, estate agents, valuers and surveyors, many of whom commenced business as long ago as the 18th century. Such broadly based firms often play an important role in the life of the town, being responsible, until recently anyway, for its cattle market and the town's salerooms – both important features of provincial towns.

They manage farms and landed estates; sell houses, farms, fields and factories.

What is now known as estate agency, that is the sale of houses, was and is just part of a general practice encompassing all aspects of property – sales, purchases, surveys, valuations and much more. Rural practices deal with the sale and purchase of agricultural land, farm valuations for revenue purposes, tenant right and so on. The older firms sometimes retain archives which record the sale and values of landed property going back two hundred years or more.

Such professional firms contribute hugely to the business life of the nation, dealing with such matters as compulsory purchase (should the State decide to drive a motorway through your property!), inheritance tax, capital gains tax, compensation claims and all the rest.

As for estate agents, this is what *The Times* said in March 2004:

"Estate agents need sympathy

However grim the waste of Hell, they must have a very good housing market. There is no group of people held to be more deserving of eternal damnation than estate agents. Fancy a flat down by the wood of adulterers or on the plain of thieves? Down on Beelzebub's high street you can hardly move for shifty-looking men clutching tape measures, their pointed tails poking through the flaps of their sharp suits.

The Office of Fair Trading (OFT) knows full well that its highly critical report on the estate agency profession, published yesterday, will gain a warm reception. The OFT reveals that one in four people who have recently sold a home is dissatisfied with the services of his/her estate agent. Agents are accused of charging too much, failing to pass on offers and creating too many delays.

Some estate agents, of course, are guilty of these crimes. But compared with the behaviour of the public the average estate agent is a model of probity. Take gazumping. It is not the estate agent who is to blame when, at the point of exchanging contracts on your dream four-bed in Carshalton, you are squeezed out by another buyer with more cash. The estate agent is legally bound to pass on offers to his client. It is the greedy, unprincipled vendor who decides to take the extra cash – and who then hides behind his estate agent, who he knows will get the blame.

Buyers are hardly any better. A good number of the people you let into your home clutching a set of estate agents' particulars are not really interested in moving house; they just happened to be out for a Sunday drive and thought they would nose around someone else's home They have come to get some ideas for doing up their bathroom or simply to snigger at the wallpaper. If estate agents seem to charge a lot of money it is hardly surprising when you consider that they have to entertain timewasters, supply them with brochures and accompany them to viewings of far-flung properties.

At the height of the dotcom boom there was a proliferation of websites which sought to cut the estate agent out of the housebuying process. People say they don't trust estate agents, but it soon became obvious how much we trust people who sell their homes privately no farther than we could fling their three-bed des res. The websites rapidly closed when it became obvious that private vendors, who are not bound by the Property Misdescription Act, tell far worse porkies than estate agents do.

It is not estate agents who cover up cracks with strategically placed pieces of furniture, who dry out damp patches with blow torches a quarter of an hour before the prospective buyers arrive. It is you and I. I certainly can't claim any moral superiority over the estate agency profession. I once sold a house bang next to a railway line by deliberately arranging viewings on the day of a rail strike.

One of the most bizarre charges laid by the OFT is that there is "little variation" in estate agents' fees – accusing them, in effect, without any evidence, of running an unconscious cartel. Yet most estate agents I know would sooner slit their rivals' throats than co-operate in a fee-fixing arrangement. Of course fees are similar; in a free market no agent can afford to charge much more than his rivals and none can afford to sell their services below cost.

On the subject of lack of competition and poor value for money the featherbedded civil servants at the OFT can hardly talk. How many OFTs are there? Er, just the one. How do we taxpayers know that we are getting value for money? We don't. If I were faced with the choice of jumping into a life-boat with an enterprising estate agent or with a stuffy bureaucrat, I know which I would choose.

Ross Clark."

Index

(of items not included alphabetically in the text)

Bibliography

The English House Through Seven Centuries. Olive Cook. Nelson.
Dictionary of Architecture. Penguin.
Old English Houses. Hugh Braun. Faber & Faber.
Stones of Britain. B.C.G. Shore. Leonard Hill (Books) Ltd.
Technical Terms for Property People. D.H. Chapman. Estates Gazette.
A Biographical Dictionary of British Architects, 1600–1840. Howard Colvin. John Murray.

Acknowledgments

The author is indebted to all whose properties feature in this book. No discourtesy is intended by not having consulted them, but contact with everyone was not possible.

The following have contributed with advice or practical assistance, and their help is acknowledged with many thanks: Mr John Hunter, FSA, Mr Roger Lord, and, of course, Carolyn Munro.